China's Military Activities in the East China Sea

Implications for Japan's Air Self-Defense Force

Edmund J. Burke, Timothy R. Heath, Jeffrey W. Hornung, Logan Ma,
Lyle J. Morris, Michael S. Chase

Prepared for the United States Air Force
Approved for public release; distribution unlimited

For more information on this publication, visit www.rand.org/t/RR2574

Library of Congress Cataloging-in-Publication Data is available for this publication.
ISBN: 978-1-9774-0098-7

Published by the RAND Corporation, Santa Monica, Calif.

Preface

This report examines Chinese maritime and air activity near Japan, especially around the Senkaku Islands. It considers implications for the Japanese Air Self-Defense Force (JASDF) and offers recommendations on ways to manage the challenges presented by Chinese air activity.

A long-standing rivalry between China and Japan has intensified in recent years, owing in part to growing parity between the two Asian great powers. Although the competition involves many issues and spans political, economic, and security domains, the dispute over the Senkaku Islands remains a focal point. This paper examines how China has stepped up its surface and air activities near Japan, in particular near the Senkaku Islands. It surveys the patterns in Chinese vessel and air activity and considers Japan's responses to date. The paper concludes that resource constraints and limited inventories of fighter aircraft impose formidable obstacles to the JASDF's ability to match Chinese air activity. Given China's quantitative advantage in fighter aircraft, Japan's current approach may not be sustainable. The paper offers recommendations for the consideration of the United States and Japan.

The research reported here was commissioned by the U.S. Air Force and conducted within the Strategy and Doctrine Program of RAND Project AIR FORCE as part of a fiscal year 2017 project on the growing reach of Chinese aerospace power. This report documents work originally shared with the U.S. Air Force in December 2017.

RAND Project AIR FORCE

RAND Project AIR FORCE (PAF), a division of the RAND Corporation, is the U.S. Air Force's federally funded research and development center for studies and analyses. PAF provides the Air Force with independent analyses of policy alternatives affecting the development, employment, combat readiness, and support of current and future air, space, and cyber forces. Research is conducted in four programs: Strategy and Doctrine; Force Modernization and Employment; Manpower, Personnel, and Training; and Resource Management. The research reported here was prepared under contract FA7014-16-D-1000.

Additional information about PAF is available on our website: www.rand.org/paf.

This report documents work originally shared with the U.S. Air Force in January 2018. The draft report, issued on January 22, 2018, was reviewed by formal peer reviewers and U.S. Air Force subject-matter experts.

Contents

Figures

Summary

A traditional, long-standing rivalry between China and Japan has intensified in recent years, owing in part to a growing parity between the two Asian great powers. The two countries compete for commercial opportunities and diplomatic influence in Asia and beyond, and they have also stepped up military modernization efforts aimed in part at one another. Moreover, the deepening competition has coincided with increasing distrust and hostility between the two countries. Despite mutual economic dependence, the leaders and peoples of both countries remain suspicious and wary of one another.

One of the most visible and vexing focal points of the rivalry has been the dispute concerning sovereignty over the Senkaku Islands (the islands are under Japan's administrative control, but Beijing also claims them and refers to them as the Diaoyu Dao). In 2012, the purchase of three of the Senkaku Islands from their private owners by Japan's national government enraged China, spurring its leaders to direct a major increase in the presence of government and fishing vessels near the disputed islands. The increased operational tempo has strained Japan's ability to match the Chinese presence. Occasionally, China has also surged the number of vessels in the area, temporarily overwhelming the Japan Coast Guard ships patrolling the area.

A similar competition has emerged in the air. Since 2012, China has dramatically increased the number of sorties and the variety of military aircraft near Japan. Fighter planes from both countries now routinely fly in close proximity to one another, raising the risk of miscalculation and dangerous crises. China's large inventory of fighter aircraft has enabled it to fly frequent missions near Japan, straining the limited resources of the Japanese Air Self-Defense Force (JASDF). Fiscal year 2016 saw the largest number of JASDF scrambles—1,168 in total—73 percent of which were against Chinese aircraft primarily flying near and around the Senkaku Island chain and the East China Sea. The stress of constantly responding to the Chinese air activities has also imposed opportunity costs for training in the Japanese fighter force, raising concerns that the demands of responding to the level of Chinese air activity could lead to the erosion of the JASDF's combat readiness. To cope with the added demand for air patrols, the JASDF has made adjustments in protocol, deployments, and acquisitions. However, given long-term trends in acquisitions and China's quantitatively superior inventory, these changes may not be sufficient to enable Japan to keep pace over the longer term.

Changes may be required over the longer term to cope with a persistent Chinese military air presence near Japan. This paper offers some recommendations toward that end, including potential greater involvement of ground-based air defense forces in tracking Chinese aircraft and allocating more aircraft to the southern sectors. More thinking may also be required to plan for contingencies in which China either provokes a crisis or directs a surge in aircraft to present the JASDF with overwhelming numbers.

Acknowledgments

The authors would like to thank the reviewers for their advice on the draft and the officials and analysts in the United States and Japan who shared their insights with us during the course of the project. In addition, the authors would like to thank Cristina Garafola, who was extremely helpful in aggregating and analyzing the data from Japan's Ministry of Defense cited in the paper.

Abbreviations

ADIZ	Air Defense Identification Zone
FY	fiscal year
GSDF	Ground Self-Defense Force (Japan)
ISR	intelligence, surveillance, and reconnaissance
JASDF	Japanese Air Self-Defense Force
MSDF	Maritime Self-Defense Force (Japan)
NDPG	National Defense Program Guidelines
PLA	People's Liberation Army
PLAAF	People's Liberation Army Air Force
SDF	Self-Defense Force (Japan)
USAF	U.S. Air Force

1. Introduction

In recent years, China and Japan have experienced a dramatic increase in nonlethal encounters between military aircraft near Japan. Chinese military aircraft have flown with increasing frequency near the Senkaku Islands and the Miyako Strait, the waterway between Miyako Island and Okinawa, which Chinese strategists regard as a critical passageway through the first island chain. The higher rate of activity has spurred Japan to adjust deployments and increase its acquisitions to keep pace with the growing Chinese presence. Although neither Tokyo nor Beijing has any interest in allowing any situation to escalate to war, the tense standoff carries a growing risk of miscalculation. Moreover, China's numerical superiority, combined with its operational initiative in this area and Japan's own operational choices in responding to Chinese aerial incursions, threatens to overwhelm the ability of the Japanese Air Self-Defense Force (JASDF) to effectively defend what Japan views as its airspace.

The increase in interactions between Chinese and Japanese military aircraft near the Senkaku Islands (Diaoyu Dao in Chinese) has occurred against a backdrop of growing competition for influence between the two great powers and a hardening of Beijing's stance regarding its sovereignty claims. Accordingly, both sides are employing military and paramilitary forces to signal resolve, demonstrate presence, and enforce sovereignty claims. Much of this competition takes place between nonmilitary surface ships, with the Chinese Coast Guard (CCG), China's maritime militia, and Chinese fishing vessels attempting to advance Chinese claims on the one side and the Japan Coast Guard tasked with the protection of Japan's position on the other. Although it includes a section on the maritime dimension, the purpose of this report is to explore the nature and progression of competition in the air domain, with an emphasis on Japan's military and security responses to the new security environment that China has forged around the Senkaku Islands, and then offer some recommendations for potential responses in a variety of domains.

The report is organized into three parts. Chapter Two provides strategic context for the increased air activity. It outlines the intensifying rivalry between the two traditional great powers of Asia and how this has exacerbated deepening perceptions of threat on both sides. This chapter also reviews the dispute over the Senkaku Islands, which remains the main flashpoint and the focus of much of the recent air activity. Chapter Three examines increased maritime patrols near the islands, especially since 2010. Chapter Four reviews the patterns in air activity between China and Japan. China has increased the number and intrusiveness of its flights, as well as the involvement of a diverse array of military aircraft in patrols near Japan. This chapter also examines the range and breadth of adjustments made by Japan in response to the increasingly intrusive presence of the Chinese aircraft, ranging from adjustments in deployments and policy to changes in acquisitions of fighter aircraft. Chapter Five draws conclusions about the likely future trajectory of interactions between the two sides. It also considers some recommendations for the United States and Japan.

2. Strategic Context

Growing tensions between China and Japan are ultimately due to a deepening competition for leadership between the two traditional Asian rivals.[1] After decades of rapid economic growth, China surpassed Japan in 2010 to become the world's second-largest economy.[2] China's military, the world's largest, has experienced rapid gains in modernization that have enabled China to close much of a perceived gap in technological sophistication between its platforms and those fielded by Japan.[3] The two countries also compete regularly for investment projects and diplomatic influence throughout the Asia-Pacific region and, increasingly, Africa and other parts of the world.[4] In 2017, China published its first white paper outlining its vision of security leadership in Asia. The paper included many statements at odds with the preferences of Japan, such as criticism of alliances and the promotion of norms and institutions based on China's "five principles of peaceful coexistence," which are a series of political principles that advocate noninterference in the internal affairs of other countries and the peaceful resolution of disputes.[5] Japan, by contrast, upholds its alliance with the United States and advocates for existing norms and institutions. When speaking about his vision of the Asia-Pacific region, Prime Minister Shinzo Abe described a new era dawning in which the region will "make freedom, human rights, and democracy our own and respect the rule of law."[6] Japan's National Security Strategy, released in 2013, provides a strategic framework for such a vision, outlining the Japanese government's support for upholding values, like freedom, democracy, fundamental human rights, and the rule of law, and expressing continuing commitment to its alliance with the United States.[7]

The intensifying competition has coincided with a hardening of Chinese attitudes regarding the country's sovereignty claims. Since his ascent to power, Xi Jinping has directed greater efforts to consolidate control over disputed territory, albeit in a manner that has avoided war and supported China's revitalization through "peaceful development."[8] President Xi has stated that

---

[1] Jeffrey Reeves, Jeffrey Hornung, and Kerry Lynn Nankivell, eds., *Chinese-Japanese Competition and the East Asian Security Complex: Vying for Influence*, New York: Routledge, 2017; Richard McGregor, *Asia's Reckoning: China, Japan, and the Fate of U.S. Power in the Pacific Century*, New York: Viking, 2017.

[2] "China Passes Japan as Second Largest Economy," *New York Times*, August 15, 2010.

[3] Reuters, "Military Strength Comparison: China, Japan, and North Korea," March 28, 2013.

[4] Yun Sun, "Rising Sino-Japanese Competition in Africa," *Brookings Institution*, August 31, 2016.

[5] State Council Information Office, People's Republic of China, "China's Policies on Asia Pacific Cooperation," January 11, 2017.

[6] Ministry of Foreign Affairs of Japan, "Address by Prime Minister Shinzo Abe at the 'Shared Values and Democracy in Asia' Symposium," January 19, 2016.

[7] Prime Minister of Japan's Office, "National Security Strategy," December 17, 2013.

[8] Xinhua, "Xi Jinping Vows Peaceful Development While Not Waiving Legitimate Rights," January 29, 2013b.

China will "absolutely not give up [its] legitimate rights and interests, and will definitely not sacrifice the state's core interests."[9] In recent years, China has taken more assertive measures to enforce its claims. It seized control of Scarborough Reef from the Philippines in 2012. Beginning around 2014, China dramatically stepped up the construction of artificial islands in the South China Sea. Media have also reported sporadic Chinese incursions along the Indian border.[10] These incursions reached a boiling point in the summer of 2017, when hundreds of Chinese and Indian troops engaged in a two-month standoff on the Doklam plateau.[11] In contrast to a long-standing emphasis on upholding regional and international stability in order to enable domestic growth, Chinese leaders thus increasingly balance the pursuit of stability with an incremental expansion of influence, albeit an expansion of a limited and opportunistic type that seeks to realize gains in the least destabilizing manner possible. This expansion of China's influence is perhaps most apparent (and publicly reported) in the militarization of some of its artificially expanded features in the South China Sea, where its claims chafe with those of other claimants—in particular, Vietnam and the United States.

In response to greater Chinese assertiveness, Japan has pushed back against Chinese actions, particularly in the maritime domain.[12] In the diplomatic realm, Japan has sought stronger strategic relationships in the Indo-Pacific region, including with Australia, India, and Southeast Asian states such as the Philippines and Vietnam. While not explicit, this networking strategy gives Japan an ever-growing base of diplomatic support to rely on when countering Chinese provocations. Some of these relationships have seen collateral improvements in operational ties as well. But Japan's most significant efforts to push back on China are in its military improvements. The government has prioritized a defense posture more focused on the region and the procurement of assets meant to strengthen the capabilities of the Japanese Self-Defense Force (SDF) in island defense. This includes new intelligence, surveillance, and reconnaissance (ISR) and antiship warfare capabilities; growth in the submarine and destroyer fleet; the establishment of an Amphibious Rapid Deployment Brigade; the stationing of SDF assets on islands close to the Senkakus, and the establishment of an air wing in Okinawa. It has also increased the Japan Coast Guard budget and established a Coast Guard patrol unit tasked specifically with patrolling

---

[9] Xinhua, "Xi Jinping Stresses at the Third Collective Study Session of the Political Bureau to Make Overall Planning for Domestic and International Situations," January 29, 2013a.

[10] Louisa Lim and Frank Langfitt, "China's Assertive Behavior Makes Neighbors Nervous," *All Things Considered*, NPR, November 2, 2012. See also Simon Denyer, "Tensions Rise Between Washington and Beijing over Man-Made Islands," *Washington Post*, May 13, 2015.

[11] Sarah Zheng, Liu Zhen, and Kristin Huang, "China 'Halts Road Building' to End Border Stand-Off," *South China Morning Post*, August 29, 2017.

[12] Jeffrey W. Hornung, "Japan's Growing Hard Hedge Against China," *Asian Security*, Vol. 10, No. 2, 2014, pp. 97–122.

the Senkaku Islands.[13] Finally, it has strengthened its ties with the United States to include the establishment of an alliance coordination mechanism meant to enhance coordination between the two states to respond to gray-zone activities during peacetime all the way up to contingencies.

Intensifying Mutual Perceptions of Threat

Of course, the current mistrust has some context: It is part and parcel of a long-lived, historical animosity that spans centuries. A key driver for the Chinese was Japan's invasion and occupation of China in the early twentieth century, which inflicted both physical destruction and human suffering on the Chinese. From China's perspective, Japan's aggression during World War II exposed and exploited China's military weaknesses, fostering a deep sense of victimization among the Chinese and leaving scars on the Chinese psyche. In spite of joint efforts to reduce and manage tensions in the post–World War II era, China harbors deep suspicions toward Japan's sincerity in owning up to its imperialist past and atrocities against China. For the Japanese, there is a sense that it has made good faith efforts at reconciliation with China, and that Beijing is simply stoking anti-Japanese nationalist sentiments to bolster the legitimacy of the Chinese Communist Party. This dichotomy has created somewhat of a security dilemma in which, despite their extensive interactions in economics and trade, the two countries are undertaking investments in military capabilities, adjusting organizational approaches, and refining their strategies and operational concepts in ways that suggest a growing security competition.

In recent years, Chinese-Japanese relations have experienced some thawing but have generally remained strained. The two sides have made tentative progress toward improved cooperation on economic issues such as the Asian Development Bank and the Asian Infrastructure Investment Bank. Japan also changed its policy in 2017, offering to involve itself in elements of the Belt and Road Initiative. However, although Presidents Xi and Shinzo Abe have met on the sidelines of several multilateral events, no bilateral state visits have taken place since 2011, reflecting the low level of trust. Tensions between China and the United States over trade, Taiwan, and maritime disputes have added to strains in the China-Japan relationship. In 2013, U.S. Secretary of State John Kerry affirmed that the United States would defend its ally Japan during a visit, which infuriated Beijing. President Donald J. Trump's embrace of Prime Minister Abe and escalating trade feud with China have spurred Beijing to adopt hard-line positions on many disputed issues. In an increasingly polarized Asia characterized by a sharpening competition for influence and status, the prospects of a China-Japan agreement over the status of the Senkaku Islands have dimmed considerably.

---

[13] See Michael Beckley, "The Emerging Military Balance in East Asia: How China's Neighbors Can Check Chinese Naval Expansion," *International Security*, Vol. 42, No. 2, Fall 2017, pp. 78–119. See also Lyle J. Morris, "The New 'Normal' in the East China Sea," *Diplomat*, February 24, 2017a.

In short, over the past ten years, the two sides have shown growing levels of distrust and hostility. Accusations and recriminations over recent controversies have fueled resentment and anger between the people of the two countries. In 2016, a Pew poll showed that between 80 and 90 percent of the public in each country holds a negative view of the other.[14] Amid strained relations, leaders in both countries thus face strong public pressure to maintain a sturdy defense of their countries' respective stances. Military officials in each country have similarly characterized the other as a threat. In 2016, the People's Liberation Army (PLA) established five joint theater commands as part of a broader reorganization.[15] The Eastern Theater command is widely understood to bear the responsibility for planning and preparing for contingencies involving Japan in the East China Sea.[16] The PLA also formed the East China Sea Joint Operations Command Center, the first of its kind anywhere on its periphery, in 2014.[17] The new center is ostensibly responsible for overseeing the PLA's day-to-day operations in the area, but little is known about its authorities or how it deconflicts or manages air and maritime space. China's 2015 defense white paper criticized Japan's military modernization as a "grave concern" for the region.[18] Japan responded by naming China as a potential threat in its own 2015 defense white paper.[19]

These factors have set the stage for a key area of competition that has emerged between the two countries: the dispute over sovereignty of the Senkaku Islands.

Senkaku Islands Dispute

China's imperative to demonstrate its status as a leading Asian power and the deepening mutual distrust between China and Japan have elevated the salience of the long-running dispute over the Senkaku Islands. The strategic value of the islands for both countries also lies in the access provided through the first island chain and their location astride Japan's sea lines of communication. But for Japan, in particular, there is also strategic value in preventing China from being able to place ISR assets close to the Japanese archipelago that would enable China to monitor both Japanese and U.S. activity in the region. Moreover, the Senkaku Islands may hold economic value. Located roughly 300 km from China's coast and 400 km from Okinawa, the

---

[14] Bruce Stokes, "Hostile Neighbors: China vs. Japan," *Pew Research Center*, September 13, 2016.

[15] See Michael S. Chase and Jeffrey Engstrom, "China's Military Reforms: An Optimistic Take," *Joint Forces Quarterly*, No. 83, 4th Quarter 2016, for an excellent overview of the reforms.

[16] Peter Wood, "Snapshot: Eastern Theater Command," *China Brief*, March 14, 2017.

[17] Strategic Research Group, "A Study on the People Liberation Army's Capabilities and Increasing Jointness in the East China Sea," Air Staff College research memo, *Air Studies*, No. 2, June 17, 2016.

[18] State Council Information Office, People's Republic of China, "China's Military Strategy," May 26, 2015.

[19] "Defense White Paper Highlights Threat Posed by China," *Japan Times*, July 21, 2015.

islands are situated close to potentially lucrative natural gas fields and fishing grounds.[20] Indeed, discoveries of potential energy deposits near the islands in the 1970s may have spurred China to establish its ownership claims. Before that period, Beijing made little effort to contest Japanese ownership of the Senkaku Islands, and, accordingly, Tokyo has refused to even acknowledge that ownership of the islands may be in dispute. Washington does not take a position regarding sovereignty of the islands, despite having administered control over them during the U.S. occupation of Okinawa and then transferred that back to Japan after Okinawa's reversion in 1972.[21] However, over the past decade, U.S. authorities have been clear in affirming that the United States recognizes Japanese administrative control, including in public statements that the islands fall under Article 5 of the U.S.-Japan Treaty of Mutual Cooperation and Security.[22]

For decades, Chinese and Japanese vessel and aircraft interactions near the Senkakus Islands were minimal. However, key developments within the past ten years have spurred both governments to significantly bolster their military and paramilitary forces in the area. In 2010, a Chinese fishing boat collided with two patrol boats of the Japan Coast Guard, touching off a diplomatic row. Chinese officials canceled numerous high-level engagements, suspended exports of rare earth minerals to Japan, and tolerated widespread mass protests and vandalism against Japanese property in China. Japan released the fishing boat captain after detaining him for over two weeks, but the incident elevated the islands as a point of friction in the China-Japan relationship.[23]

Bilateral strains deepened considerably after Japan's national government purchased three of the islands from a private Japanese owner in 2012, ostensibly in an effort to preempt the purchase and development of the islands by Shintaro Ishihara, at the time the outspoken nationalist governor of Tokyo. Protests, some of them violent, spread through China. In recent years, Chinese authorities have employed a broad array of measures that amount to a hybrid warfare (briefly, intimidation short of warfare) of the sort used elsewhere in the region by China to bolster its position regarding the Senkakus.[24] In addition to regular media fusillades and diplomatic protests, Chinese officials have employed economic tools of coercion to compel Japanese concessions, including reductions in the number of Chinese tourists to Japan and the

---

[20] Ministry of Foreign Affairs of Japan, "Senkaku Islands," April 15, 2014.

[21] Despite the reversion of the islands back to Japan in 1972, two of the islands, Kuba Island and Taisho Island, remain leased to the United States per the U.S.-Japan Status of Forces Agreement. In the past, the U.S. military has used these islands as bombing ranges.

[22] Philip Crowley, "Remarks to the Press," U.S. State Department, September 23, 2010; Justin McCurry and Tania Branigan, "Obama Says US Will Defend Japan in Island Dispute with China," *Guardian*, April 24, 2014.

[23] "Japan Frees Chinese Boat Captain amid Diplomatic Row," *BBC*, September 24, 2010.

[24] For a discussion of how Beijing interprets international norms and agreements in ways that serve its political and strategic goals in the East China Sea and vis-à-vis Japan, see Edmund J. Burke and Astrid Stuth Cevallos, *In Line or Out of Order? China's Approach to ADIZ in Theory and Practice*, Santa Monica, Calif.: RAND Corporation, RR-2055-AF, 2017.

aforementioned curtailment of exports of rare earth minerals to Japan after the 2010 fishing boat incident.[25] Beijing has also advanced more elaborate legal arguments for its ownership claims, which Japan has summarily dismissed.[26] It is this post-2012 time frame that is the main focus of our analysis, starting with a brief overview of the increasing maritime vessel encounters between Japan and China around the Senkaku Islands.

---

[25] Kyung Lah, "Japan Hopes for U.S. Help in Row with China," *CNN*, November 13, 2010.

[26] State Council Information Office, People's Republic of China, "Diaoyu Dao, an Inherent Territory of China," August 23, 2014.

3. Chinese-Japanese Maritime Vessel Interactions

Given the location of the Senkaku Islands and their association with territorial integrity and national pride, the 2012 purchase of the islands by Japan from a private Japanese citizen was the catalyst that allowed simmering tensions between China and Japan to spill out into the open. On China's side, there has been a marked increase in the involvement of maritime law enforcement, as well as maritime militia and fishing vessels near the disputed islands, since 2012. Beijing has dramatically increased the level of intrusions into the 24-nautical-mile (nm) contiguous zone and 12 nm territorial sea around the Senkaku Islands. In the days following Tokyo's purchase of the islands in 2012, for example, two China Marine Surveillance vessels penetrated the territorial seas of the Senkaku Islands, setting a precedent for an intensified penetration campaign into the waters around the Senkakus by Chinese maritime law enforcement vessels.[1] By the end of 2012, the Japan Coast Guard reported that Chinese Coast Guard ships had intruded into Senkaku territorial waters 68 times since September 11, an unprecedented number of intrusions. The campaign continued, with 188 vessels penetrating the territorial sea in 2013, 88 in 2014, 86 in 2015, and 121 in 2016.[2] Since mid-2014, on average, Chinese government vessels have penetrated the territorial seas seven to nine times a month and have carried out 70–90 incursions in the contiguous zone—waters within which states can exercise enhanced jurisdiction but that do not constitute sovereign territory like those of a territorial sea.[3] Over this period of time, Chinese Coast Guard behavior followed a relatively consistent pattern, with vessel transits into the territorial seas usually lasting a few hours and Chinese Coast Guard officers maintaining communication with Japan Coast Guard officers to avoid miscalculations. There have been very few instances of Chinese Coast Guard vessels undertaking actions that may be considered a violation of laws of innocent passage, which may compel Japan Coast Guard officers to adopt more aggressive measures in response.[4] In other words, Chinese vessels appear intent on demonstrating that they can exercise administrative control over the Senkakus by establishing a coast guard presence in territorial waters, but in a manner that would not escalate to a military conflict with Japan.

Since 2015, the Chinese Coast Guard has started to deploy armed cutters, including a 76 mm autocannon on larger ships, in a shift from previous practice.[5] The presence of Chinese maritime

---

[1] Morris, 2017a.

[2] Ministry of Foreign Affairs of Japan, "Trends in Chinese Government and Other Vessels in the Waters Surrounding the Senkaku Islands, and Japan's Response," June 30, 2017.

[3] Ministry of Foreign Affairs of Japan, 2017.

[4] For the definition of *innocent passage*, see the United Nations Convention on the Law of the Sea, Part 2, Territorial Sea and Contiguous Zone, Article 19, Meaning of Passage, December 10, 1982. See also Morris, 2017a.

[5] Morris, 2017a.

militia—fishermen and crew of civilian ships who receive military training or are under some type of military command and control arrangement—near the Senkakus adds further complexity to the security environment in the East China Sea. Various analysts have described such nonmilitary, coercive actions as "gray zone" or "hybrid" tactics designed to undermine Japan's position and strengthen de facto Chinese control without provoking a war. The success of gray-zone tactics on the water has allowed China to revise elements of the status quo while depriving the U.S.-Japan alliance of a casus belli for military action, leading at least some to doubt Washington's resolve should China cross the threshold into open conflict.[6]

China has also deployed its fishing vessels in a choreographed manner to further Chinese aims at asserting sovereignty over the Senkakus. For example, in several instances, China has sent large numbers of fishing vessels in an attempt to overwhelm Japan's ability to respond to them. From August 5 to August 9, 2016, for example, between 200 and 300 Chinese fishing vessels and 15 Chinese Coast Guard vessels entered and loitered inside the contiguous zone around the islands, stressing the Japan Coast Guard's capacity to manage and react to the large number of vessels from China.[7] Such a "swarm" tactic, if adopted in the future, will undoubtedly present a serious challenge for the Japan Coast Guard.

In response to these challenges, Japan's SDF has made efforts to better defend against Chinese maritime threats. The most visible change has been the creation of a small amphibious unit within the Ground Self-Defense Force (GSDF). In March 2018, the GSDF launched a 2,100-member Amphibious Rapid Deployment Brigade trained for island defense.[8] Once completed, at an unspecified future date, its total strength will reach approximately 3,400.[9] Japan has also expanded its inventory of naval platforms. The Maritime Self-Defense Force (MSDF) maintains 46 destroyers, 4 of which are currently Aegis equipped, and 17 submarines.[10] The service plans to increase the destroyer fleet to 54 (among these, its Aegis-equipped destroyers will be increased to 8) and the submarine fleet to 22. This will include 2 smaller, 3,900-ton

---

[6] James Holmes and Toshi Yoshihara, "Deterring China in the 'Gray Zone': Lessons of the South China Sea for U.S. Alliances," *Orbis*, Vol. 61, No. 3, Summer 2017, pp. 322–339. For an excellent discussion of how these tactics can be understood as deliberate attempts to manipulate Japanese and U.S. intervention thresholds, see Ben Connable, Jason H. Campbell, and Dan Madden, *Stretching and Exploiting Thresholds for High-Order War: How Russia, China, and Iran Are Eroding American Influence Using Time-Tested Measures Short of War*, Santa Monica, Calif.: RAND Corporation, RR-1003-A, 2016.

[7] "Tokyo Trying to Draw Attention to Mass China Ship Incursions off Senkakus," *Japan Times*, April 17, 2016.

[8] Ministry of Defense of Japan, "Bōuei Daijin Rinji Kisha Kaiken Gaiyō" ["Summary of the Minister of Defense's Special Press Conference"], n.d.a.

[9] Koichi Isobe, "The Amphibious Operations Brigade," *Marine Corps Gazette*, Vol. 101, No. 2, February 2017, pp. 24–29; anonymous retired Ground Self-Defense Force officer, correspondence with Jeffrey Hornung, September 27, 2017.

[10] Ministry of Defense of Japan, "Reference 10," in *Nihon no Bōei Heisei 29 Nenban* [*Defense of Japan 2017*], Tokyo: Ministry of Defense, 2017b, p. 480.

destroyers of a new destroyer class and the construction of a new class of submarine with higher detection capabilities to conduct intelligence-gathering and surveillance activities.

Importantly, Japan has dedicated more resources to the Japan Coast Guard, which is the lead agency responding to Chinese incursions due to Japan's viewing the issue as a law enforcement effort. The Japan Coast Guard's budget has grown in recent years, with an increasing amount dedicated solely to Senkaku Island defense. In its fiscal year (FY) 2017 budget, out of $1.87 billion USD, 27 percent ($510 million USD) was earmarked for improving surveillance in the waters near the islands.[11] The Japan Coast Guard has used these resources to acquire more patrol ships, add more personnel, and establish a real-time video transmission directly to the prime minister's office. In 2016, the Japan Coast Guard also created a 12-ship unit tasked exclusively with guarding the waters off the Senkaku Islands. The unit, based on Ishigaki, consists of ten new patrol ships and two—equipped with helipads—that were transferred from other parts of Japan.[12] The office on nearby Miyako is also experiencing growth. Since the end of 2015, not only has the status of the Japan Coast Guard office been upgraded, but its personnel have also grown from 55 to 180.[13] Significantly, by 2019, the Japan Coast Guard is planning to open its first shooting range not located on the main Japanese island of Honshu. It is worth noting, however, that the Chinese Coast Guard now boasts the largest coast guard in the world in terms of total tonnage and manpower, and it could theoretically overmatch Japan Coast Guard forces in the East China Sea if it chose to concentrate its forces in that area of operations.[14]

Although much of the international attention devoted to the China-Japan rivalry and the Senkaku Islands dispute has focused on these developments, the growing level of activity and friction between the two sides has not been confined to the maritime domain. Indeed, activity in the air has also been increasing, and this is the subject of the next chapter of this report.

---

[11] "Japan Coast Guard to Spend 27% of Budget on Boosting Senkaku Surveillance in 2017," *Kyodo News*, December 22, 2016.

[12] Morris, 2017a.

[13] Ankit Panda, "East China Sea: Japan Coast Guard Plans Miyako Island Facility Upgrades," *Diplomat*, September 24, 2017.

[14] Lyle J. Morris, "Blunt Defenders of Sovereignty: The Rise of Coast Guards in East and Southeast Asia," *Naval War College Review*, Vol. 70, No. 2, Spring 2017b, p. 78.

4. Chinese-Japanese Military Aircraft Interactions

In recent years, China has dramatically increased the scale and complexity of its air activities in the East China Sea, stressing Japan's ability to respond. The intensifying strategic rivalry between China and Japan, hardening perceptions of threat and hostile intent, and the intensifying standoff over the Senkaku Islands provide the backdrop to this competition in the air. Indeed, in many ways, the interaction between the two countries military aircraft serves as a symptom of their broader strategic competition and feud over the Senkaku Islands.

The patterns in Chinese aircraft activity have exhibited several features. First, the level of Chinese air activity to which Japan must respond has increased dramatically. In January 2013, China's Ministry of National Defense publicly acknowledged that Chinese aircraft regularly conducted surveillance activities and combat air patrols in the East China Sea.[1] Although there are no publicly available data that portray the exact level of Chinese air activity in the area, a considerable increase in Chinese air activity is clearly reflected by the growing number of scrambles the JASDF has mounted in response, which is reported publicly in the aggregate but without much specificity by Japan's Ministry of Defense (Figure 4.1).[2]

Figure 4.1. Aggregate Reporting on Japanese Scrambles

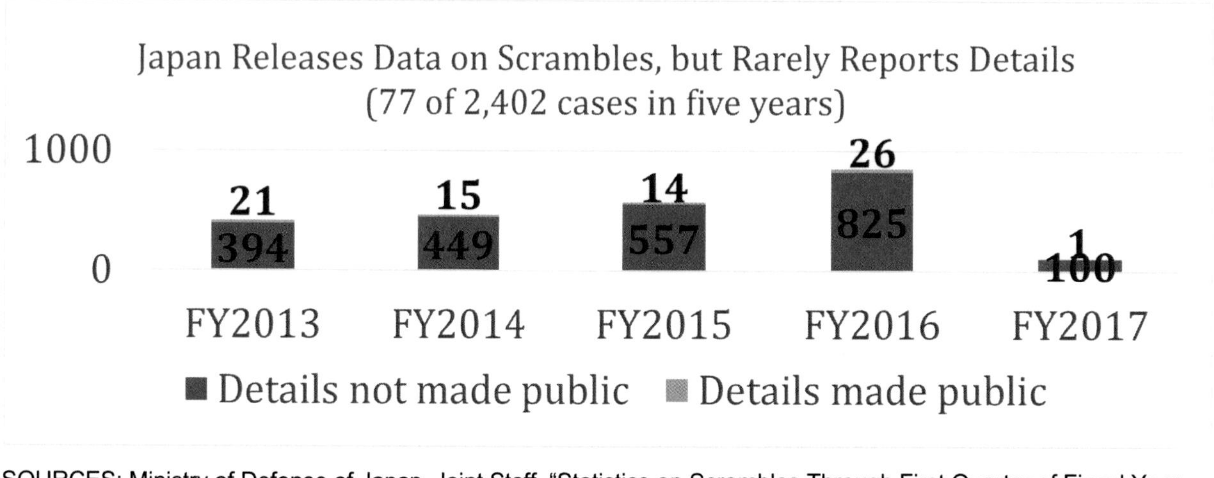

Japan Releases Data on Scrambles, but Rarely Reports Details
(77 of 2,402 cases in five years)

	FY2013	FY2014	FY2015	FY2016	FY2017
Details not made public	394	449	557	825	100
Details made public	21	15	14	26	1

SOURCES: Ministry of Defense of Japan, Joint Staff, "Statistics on Scrambles Through First Quarter of Fiscal Year 2017," press release, July 14, 2017b; Ministry of Defense of Japan, Joint Staff, "Statistics on Scrambles Through Fiscal Year 2016," press release, April 13, 2017; Ministry of Defense of Japan, Joint Staff, "Statistics on Scrambles Through Fiscal Year 2015," press release, April 22, 2015; Ministry of Defense of Japan, Joint Staff, "Statistics on Scrambles Through Fiscal Year 2014," press release, May 22, 2015; Ministry of Defense of Japan, Joint Staff, "Statistics on Scrambles Through Fiscal Year 2013," press release, April 23, 2014.

---

[1] Ministry of Defense of Japan, "China's Activities Surrounding Japan's Airspace," n.d.b.

[2] Ministry of Defense of Japan, Joint Staff, "Statistics on Scrambles Through Fiscal Year 2016," press release, April 13, 2017a.

Additionally, Chinese aircraft have intruded into Japanese airspace—both near the Senkakus and near other islands of Japan. On December 13, 2012, a Y-12 aircraft belonging to the State Oceanic Administration, a Chinese state entity, made the first-ever unauthorized intrusion into Japanese-claimed airspace in the then-45-year history of record keeping on these issues, passing by the Senkaku Islands.[3]

Second, a growing variety of Chinese combat aircraft are flying at longer ranges near Japan than had done so previously. In July 2013, China conducted its first military flight through the Miyako Strait, dispatching a Y-8 early warning aircraft through the strategically important waterway situated between the main island of Okinawa and Miyako Island. Its waters sit astride one of the few international routes through which Chinese air and naval forces are able to break past the confines of the first island chain and into the western Pacific. Two months later, China conducted its first long-range bomber flights through the Miyako Strait, with two PLA Naval Aviation H-6 bombers flying beyond the waterway before returning to the East China Sea.[4]

Third, China has also increased the complexity of operations in the area, featuring more mission-focused packages of command and control, specialized mission, fighter, and bomber aircraft. For example, the People's Liberation Army Air Force (PLAAF) conducted its first long-range bomber flights into the western Pacific on May 21, 2015, when a detachment of two PLAAF H-6K bombers flew through the Miyako Strait before retracing their flight path back to the East China Sea.[5] Previously, only People's Liberation Army Navy aviation forces flying older variants of the H-6 engaged in long-range bomber flights into the western Pacific. On September 25, 2016, the PLAAF dispatched more than 40 aircraft of various types, including H-6K bombers, Su-30 fighters, and air tankers, through the Miyako Strait and into the western Pacific to test their combat capabilities on the high seas.[6] Information publicized by Japan's Ministry of Defense also indicated the presence of intelligence-gathering aircraft in the fleet.[7]

China has justified the air activity as measures designed to monitor sovereign areas within its jurisdiction. China's 2013 defense white paper acknowledged "air vigilance and patrols at sea," explaining that they serve the purposes of "conducting counter-reconnaissance in border areas and verifying abnormal and unidentified air situations."[8] To back its claim, China has also carried out administrative measures. In November 2013, without prior formal consultation

---

[3] Ministry of Defense of Japan, n.d.b.

[4] Ministry of Defense of Japan, n.d.b.

[5] Ministry of Defense of Japan, Joint Staff, "On the Flight of Chinese Aircraft in the East China Sea," press release, May 21, 2015a.

[6] Xinhua, "China Air Force Conducts West Pacific Drill, Patrols ADIZ," September 25, 2016.

[7] Ministry of Defense of Japan, Joint Staff, 2015a.

[8] State Council Information Office, People's Republic of China, "The Diversified Employment of China's Armed Forces," April 16, 2013.

with neighboring countries, China announced the formation of an East China Sea Air Defense Identification Zone (ADIZ). The zone stretches into the East China Sea, covers the disputed Senkaku Islands, and overlaps with the ADIZs of Japan and other neighboring countries (Figure 4.2). According to a Chinese Ministry of National Defense statement, all aircraft entering the zone must identify themselves to Chinese authorities and are subject to emergency military measures should they fail to abide by the rules governing the ADIZ. The move drew immediate protest from Japan, with its Ministry of Foreign Affairs releasing a statement expressing "deep concern about China's establishment of such zone and obliging its own rules within the zone," which it described as "profoundly dangerous acts that unilaterally change the status quo in the East China Sea, escalating the situation, and that may cause unintended consequences in the East China Sea."[9]

Figure 4.2. China's Air Defense Identification Zone

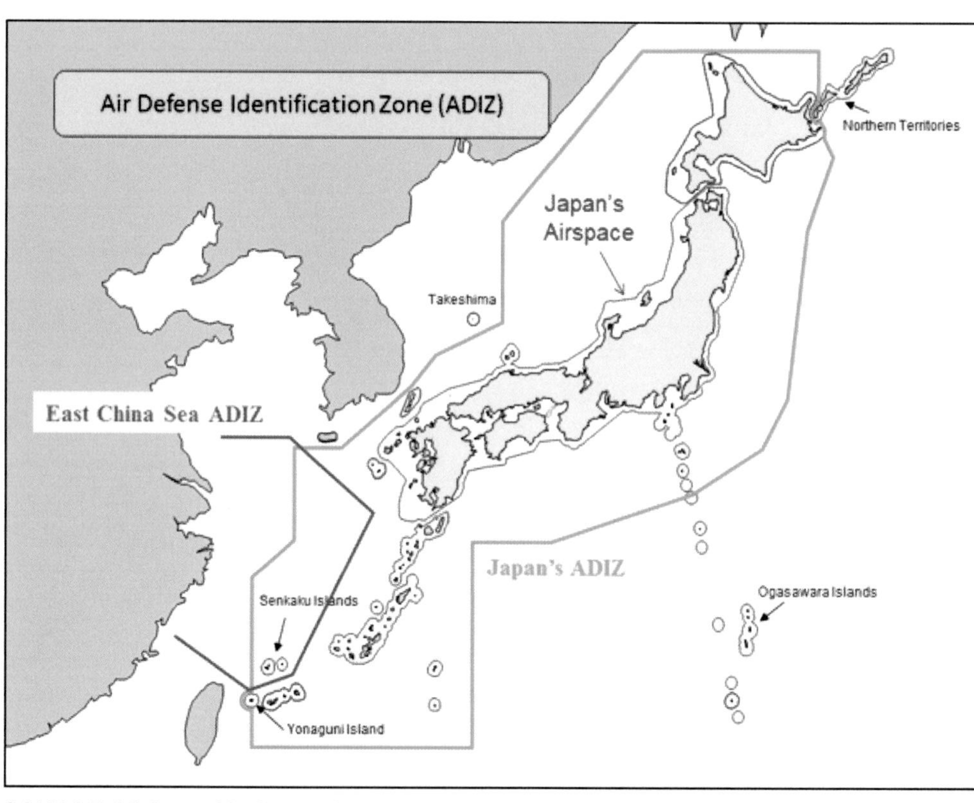

SOURCE: Ministry of Defense of Japan website.

[9] Ministry of Foreign Affairs of Japan, "Statement by the Minister for Foreign Affairs on the Announcement on the 'East China Sea Air Defense Identification Zone' by the Ministry of National Defense of the People's Republic of China," November 24, 2013.

Whatever the ostensible purpose, the rapid rise in Chinese air activity serves several tactical and strategic functions. On the tactical level, these flights offer invaluable training experiences for China's pilots. The PLAAF in particular is undergoing a shift from its traditional focus on territorial defense to an emphasis on a broader set of responsibilities, such as conventional deterrence and offensive missions.[10] To that end, flights in the East China Sea put those aspirations into practice by exposing pilots to new operational environments. Chinese air activities in the East China Sea also provide a means of gathering valuable intelligence on the disposition of Japan's defenses in the region. Information related to scrambling response times, command and control practices, and other matters is most certainly noted in interactions with opposing aircraft.

Beyond tactical functions, the air activities play a strategic role in the Sino-Japanese rivalry. China's probing actions may be viewed as an effort to draw Japanese responses that would allow China to depict Japan as an aggressor, thereby providing it with a convenient diplomatic pretext for taking more escalatory or coercive measures. In conjunction with the actions of China's maritime militia and law enforcement elements on the surface, the flights also challenge Japan's administrative control of the Senkaku Islands. Indeed, the November 2013 formation of the East China Sea ADIZ aided in facilitating that aim. Although China's ADIZ, like those of other countries, in practice lacks the force of a binding international agreement, even the pretense of legitimacy constitutes a challenge to Japan's control of the islands. Finally, in expanding its air operations in the region, China seeks to normalize its military presence in the East China Sea. For example, in the aftermath of one flight, a Chinese Ministry of National Defense spokesperson retorted that "the parties concerned" should "[get] used to such drills."[11] By constructing a new status quo, China, on the one hand, affirms the staying power of its military presence and, on the other, seeks to overtake Japan as the dominant power in the region.

Japan's Response to Chinese Air Activity

Japan has pursued numerous measures to counter China's increased air activity, including defense policy modifications, increased defense budgets, and service organizational changes. These changes began with revisions to Japan's National Defense Program Guidelines (NDPG) in December 2010. Under the notion of a "dynamic defense force," Japan shifted its focus from its traditional policy of a passive, deterrent, "basic defense force" to one focused on "readiness, mobility, flexibility, sustainability, and versatility."[12] Importantly, the NDPG shifted away from

---

[10] For more on this subject, see Michael S. Chase and Cristina L. Garafola, "China's Search for a 'Strategic Air Force,'" *Journal of Strategic Studies*, Vol. 39, No. 1, 2016, pp. 4–28.

[11] Xinhua, "China Air Force in West Pacific Drill," May 21, 2015.

[12] Ministry of Defense of Japan, "National Defense Program Guidelines for FY2011 and Beyond," December 17, 2010, p. 7.

Japan's traditional Cold War–era focus on a potential northern Soviet invasion to a focus on protecting Japan's southwestern islands from Chinese aggression. The NDPG was revised by the Abe administration in 2013 to emphasize the jointness of forces under the notion of a "dynamic joint defense force." In support of this reorientation, since coming into power in 2012, the Abe administration has increased defense spending annually for the purpose of acquiring the necessary capabilities for defense of the southwestern island chain. Since sinking to a low of $41 billion USD in FY 2012, Japan's defense budgets have slowly climbed to their current $43 billion USD in FY 2017—an absolute increase of just 5 percent in that time frame.[13]

For much of its existence, the JASDF was organized primarily to meet the Soviet and later Russian threat. As a consequence, the JASDF maintained a weaker force posture in the southwest. However, with China's growing assertiveness in the East China Sea changing the security dynamics of the region, Japan has now turned greater attention to bolstering its southwestern border defense capabilities. While the Japan Coast Guard handles policing the action of the seas, the JASDF is the only organization capable of policing Japanese airspace. The sheer volume of Chinese air activity in the region has also forced the JASDF to adjust its posture and capabilities in response. In April 2014, the JASDF reallocated aircraft to check Chinese activity by establishing the 603rd Squadron at Naha Air Base. The squadron consists of four E-2C Hawkeye airborne early warning aircraft and 130 personnel transferred from the Misawa-based 601st Squadron. Previously, the JADSF deployed aircraft from Misawa to Okinawa on a rotational basis.[14] The permanent presence of early warning aircraft in the southwestern region supplements existing ground-based radar stations, enhancing Japan's surveillance capabilities in the region.

Organizational changes were implemented as well. Until 2016, the JASDF divided the country's air defense into three main air areas: northern, central, and western. In each of these three defense areas, the JASDF maintains two air wings. In the Northern Air Area, this includes the 2nd Air Wing at Chitose and the 3rd Air Wing at Misawa. In the Central Air Area, this includes the 6th Air Wing at Komatsu and the 7th Air Wing at Hyakuri. In the Western Air Area, this includes the 5th Air Wing at Nyutabaru and the 8th Air Wing at Tsuiki. Each wing consists of two squadrons. In addition to these three main defense areas, Japan also had the Southwestern Composite Air Division, a smaller area that consisted of the 83rd Air Wing's 204th Fighter Squadron, which was solely responsible for scrambling against Chinese incursions in Japan's southwestern islands. On January 31, 2016, the JASDF reformed the 83rd Air Wing at Naha into the 9th Air Wing by supplementing the 204th Squadron with the 304th Squadron of F-15s from the 8th Air Wing at Tsuiki Air Base, thereby doubling the number of F-15s stationed in Naha.[15]

---

[13] Ministry of Defense of Japan, *Defense Programs and Budget of Japan FY2017*, Tokyo: Ministry of Defense, August 2016.

[14] David Donald, "Japan Bolsters Southern Air Defenses," *Aviation International News*, April 22, 2014.

[15] "Establishment of the New 9th Air Wing," *Japan Defense Focus*, March 2016.

This was the first new air wing since the 8th Air Wing was established in 1964, and its addition reflected a major reshuffling of combat units in the face of growing concerns regarding Japan's security environment in its southwestern airspace. The move brought the total number of fighters in the two 9th Air Wing squadrons to 40, with about 1,500 personnel.[16] On July 1, 2017, the Southwestern Composite Air Division was elevated to become the Southwestern Air Area.[17] Unlike Japan's other three defense air areas, the southwestern area consists of only one wing (the 9th Air Wing) but, like the other wings, is made up of two squadrons. This transition marked the first realignment of the JASDF since 1973.[18]

Finally, the JASDF is expanding its inventory of aircraft to meet China's increasing aerial presence. Currently, the JASDF maintains an arsenal of 201 F-15J/DJs, 52 F-4EJs, and 92 F-2A/Bs.[19] In 2017, the JASDF received its first delivery of 4 F-35As, with 6 more requested in the FY 2018 budget as part of its plan to acquire a fleet of 42 to replace the fleet of F-4EJs. In the summer of 2018, the Ministry of Defense is expected to decide on a fifth-generation, twin-engine stealth aircraft to replace the F-2 fleet when the aircraft start retiring in the 2030s.[20] For ISR, in addition to the E-2C squadron that was established in Okinawa in 2014 to patrol the southwestern island region[21] and improve the E-767 fleet's warning and surveillance capabilities, Japan looks to acquire two E-2D Advanced Hawkeye airborne early warning and surveillance aircraft, as well as three RQ-4B Global Hawk unmanned aerial vehicles to enhance persistent wide area surveillance capabilities. For logistical support, Japan seeks to add KC-46A aircraft to its current KC-767 fleet of four aerial refueling planes and add aerial refueling functions to some KC-130K and C-130H planes. For airlift, the JASDF is procuring 11 indigenously produced C-2 cargo planes, which have a more enhanced cruising range and can carry a heavier payload than the existing C-1 fleet.[22] The GSDF is supplementing these efforts with a fleet of 17 V-22 tilt-rotor aircraft, primarily for amphibious operations.

Japan's Ministry of Defense regularly publicizes information on the frequency of Chinese air activities (Figure 4.3). Most publications entail an accounting of the number of scrambles conducted in the fiscal year, with breakdowns by region and country. In circumstances deemed unusual, such as flights through the Miyako Strait, the Ministry of Defense delves into greater detail, publishing press releases detailing flight paths and the number and types of aircraft

---

[16] Matthew Burke and Chiyomi Sumida, "Japan Bolstering Southwestern Defenses," *Stars and Stripes*, January 29, 2016.

[17] There are different variants of the name in English, but in Japanese, these defense areas' titles are their geographic position (e.g., northern, central, western, or southwestern) plus *air area*.

[18] "Establishment of the Southwestern Defense Force," *Japan Defense Focus*, August 2017.

[19] Ministry of Defense of Japan, 2017b.

[20] Franz-Stefan Gady, "Japan's Air Force to Receive 100 New Stealth Fighter Jets," *Diplomat*, July 7, 2016.

[21] Kosuke Takahashi, "JASDF Forms New AEW Squadron in Okinawa," *Jane's Defence Weekly*, April 14, 2014.

[22] Ministry of Defense of Japan, *Medium Term Defense Program (FY2014–FY2018)*, December 17, 2013, p. 33.

Figure 4.3. Aggregate Reporting on Japanese Scrambles to Chinese Aircraft

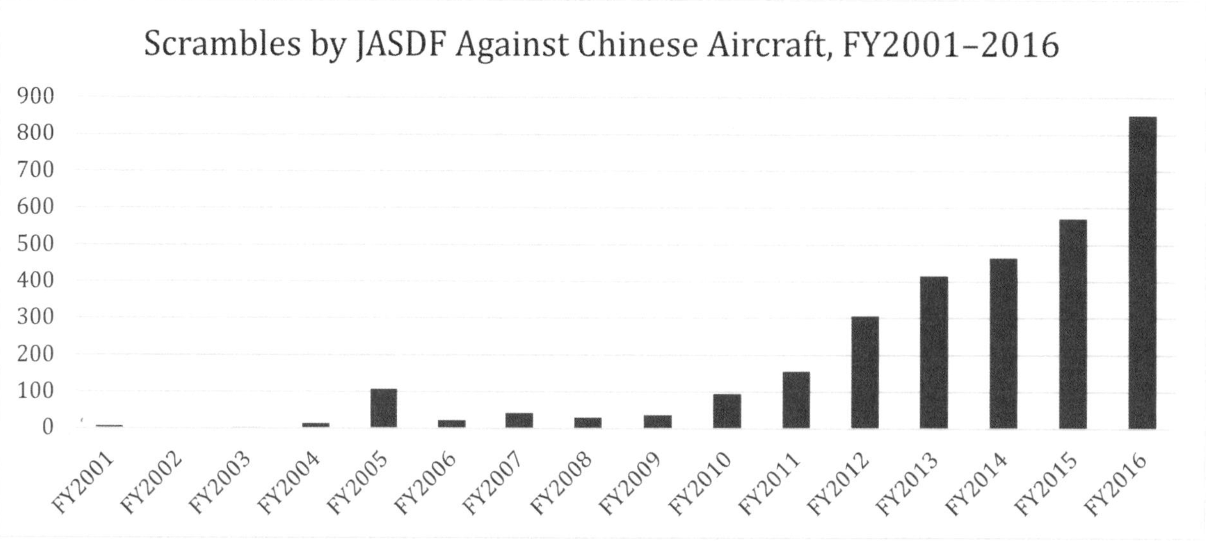

Scrambles by JASDF Against Chinese Aircraft, FY2001–2016

SOURCE: Ministry of Defense of Japan, "China's Activities Surrounding Japan's Airspace," n.d.b.

involved, as well as providing visual documentation. The data themselves represent the number of scrambles carried out by the JASDF, rather than the actual number of incursions into Japan's ADIZ. Nevertheless, the number of scrambles initiated should correlate positively with the actual number of incursions because the JASDF is expected to respond to each incursion by China.

According to a firsthand account of JADSF scrambles, once surveillance aircraft and ground-based radar detect suspicious aircraft potentially intruding on Japan's airspace, the JADSF scrambles fighters within five minutes.[23] These fighters first approach the suspicious aircraft, confirm its situation, and monitor its movements. Should the aircraft enter Japanese territorial airspace—the airspace above Japan's territorial waters—it is asked to leave or forced to land at a nearby airport. In cases of self-defense or averting present danger, the use of weapons is permitted as part of necessary actions against an intruder. With both sides flying fighter aircraft close to each other, the risk of an incident or potential for miscalculation has increased. In January 2013, Japan and China dispatched fighter aircraft in response to each other for the first time, with the PLA scrambling two J-10 fighters after the JASDF deployed F-15 fighters to observe a Y-8 early warning aircraft flying over oil and gas fields in the East China Sea.[24]

Since 2008, JASDF responses overall have increased approximately 20-fold, with the majority of scrambles directed toward Chinese flights. At the end of FY 2012, Japan's Ministry of Defense reported 306 scrambles in response to Chinese air activities. By 2016, 851 scrambles had

---

[23] "Japan: ASDF, Chinese Fighter Jets Waging Hidden Battles over East China Sea," *Sentaku*, June 1, 2017.

[24] Ian E. Rinehart and Bart Elias, *China's Air Defense Identification Zone (ADIZ)*, Washington, D.C.: Congressional Research Service, R43894, January 30, 2015.

occurred—the most scrambles on record in response to Chinese air activities.[25] According to the JASDF, its fighter aircraft performed a total of 1,168 scrambles in FY 2016 against all countries, an approximately 34 percent (295) increase compared to the previous year.[26] This represents the largest number of scrambles conducted by the JADSF since record keeping began in 1958, superseding the previous high of 944—a figure recorded during the height of the Cold War.[27] Of these 1,168 scrambles, 73 percent were in response to approaching Chinese aircraft, an increase of 49 percent from the previous fiscal year.[28] By comparison, the JADSF conducted 26 percent of scrambles (301 total) in response to Russian air activities, and 1 percent (16 total) in response to the air activities of other countries. Of the 851 scrambles against Chinese aircraft, the JADSF publicized 26 flights deemed "unusual," such as round-trip flights beyond the Miyako Strait and into the western Pacific, as well as round-trip flights from the East China Sea to the Sea of Japan.[29]

Regionally, the Southwestern Composite Air Division—since upgraded to the Southwestern Air Defense Force in July 2017—conducted the majority of these scrambles.[30] The Southwestern Air Defense Force oversees an area encompassing Japan's southwestern airspace, including the airspace above Okinawa, the disputed Senkaku Islands, and the strategically important Miyako Strait. (See Figure 4.4 for a map of Japan's air defense zones.) In FY 2016, the JASDF's Northern Air Defense Force conducted 265 scrambles, presumably mostly if not completely in response to Russian air activity, while its Central and Western Air Defense Forces conducted 34 and 66 scrambles, respectively. Meanwhile, the Southwestern Composite Air Division accounted for a disproportionate 803 scrambles, or approximately 69 percent of total scrambles conducted. Compared to the previous fiscal year, the number of scrambles conducted by the Southwestern Composite Air Division and the Northern Air Defense Force increased, while the number of scrambles conducted by the Central and Western Air Defense Forces decreased.

The first half of Japan's FY 2017—the period between April 1, 2017, and September 30, 2017—saw an unusual drop in Chinese air activity near Japanese airspace, with 287 scrambles conducted in response to Chinese air activity, compared to 407 during the same period in FY 2016.[31] In contrast, scrambles conducted in response to Russian aircraft climbed from

---

[25] Ministry of Defense of Japan, "Overview of Situation Surrounding Japan: Statistics on Scrambles Through FY 2016," slide handout, Center for Air Power Strategic Studies, May 8, 2017a.

[26] In Japan, the fiscal year begins on April 1 and ends on March 31 of the following calendar year.

[27] Japan continues to scramble against Russian aircraft in its north. In FY 2016, JASDF fighter jets scrambled 301 times against Russian aircraft, an increase of 13 times compared to the previous fiscal year. Ministry of Defense of Japan, Joint Staff, 2017a.

[28] Ministry of Defense of Japan, 2017a.

[29] Ministry of Defense of Japan, 2017a.

[30] Ministry of Defense of Japan, 2017a.

[31] Ministry of Defense of Japan, Joint Staff, "Statistics on Scrambles During the First Half of FY2017," press release, October 13, 2017c.

Figure 4.4. Japan's Air Defense Zones

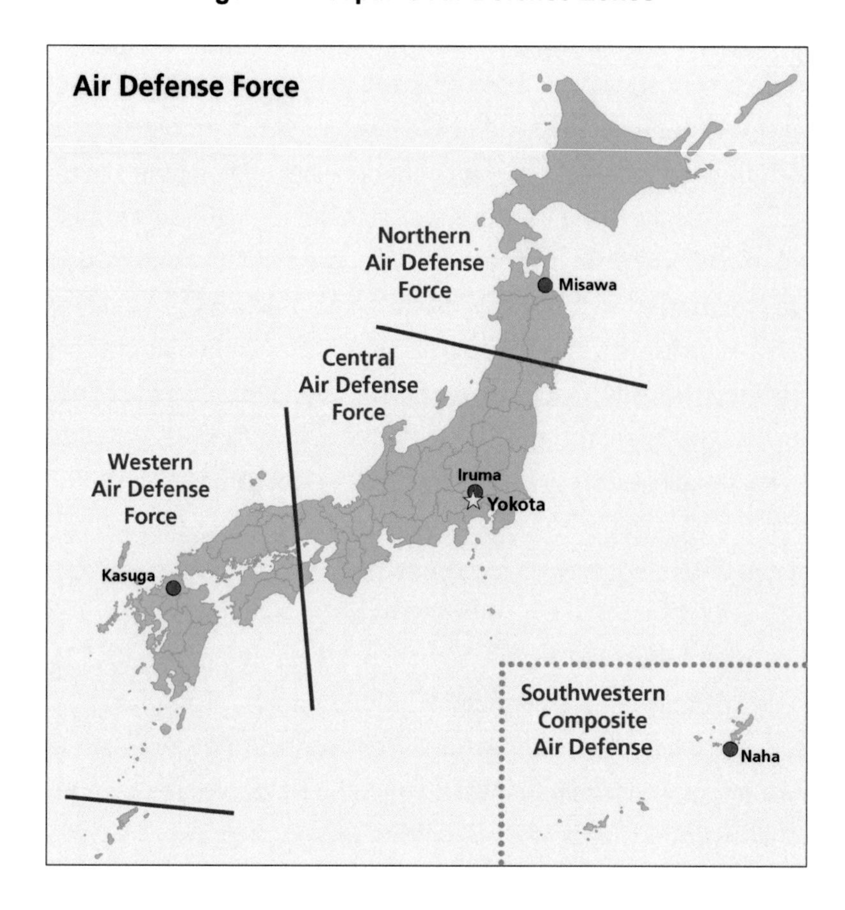

180 to 267 in the first half of the current fiscal year. Though this is speculative, Japanese interlocutors believe the reduction in Chinese flight activity can be attributed to China's desire to avoid ratcheting tensions ahead of the 19th National Congress of the Communist Party of China, which was held in Beijing in October 2017. Interestingly, this appears to have coincided with a more cautious approach than occurred over the previous seven years (see Figure 4.5) in terms of Chinese fishing activity around the Senkaku Islands during at least part of this period of time as well, according to a recent report by the Asia Maritime Transparency Initiative at the Center for Strategic and International Studies.[32] For the first half of FY 2017, however, Japan's Ministry of Defense also publicized 14 cases deemed unusual, including a case in which a Chinese drone-like object entered Japanese airspace around the Senkaku Islands (see Figure 4.6).[33]

---

[32] The assessment of the air activity is based on a discussion with Japanese defense officials in Washington, D.C., in September 2017. For further details on the fishing activity, see Asia Maritime Transparency Initiative, "Smooth Sailing for East China Sea Fishing," November 30, 2017.

[33] Ministry of Defense of Japan, Joint Staff, "Statistics on Scrambles Through First Quarter of Fiscal Year 2017," press release, July 14, 2017b.

Figure 4.5. Japanese Scrambles by Ingressing Aircraft Origin

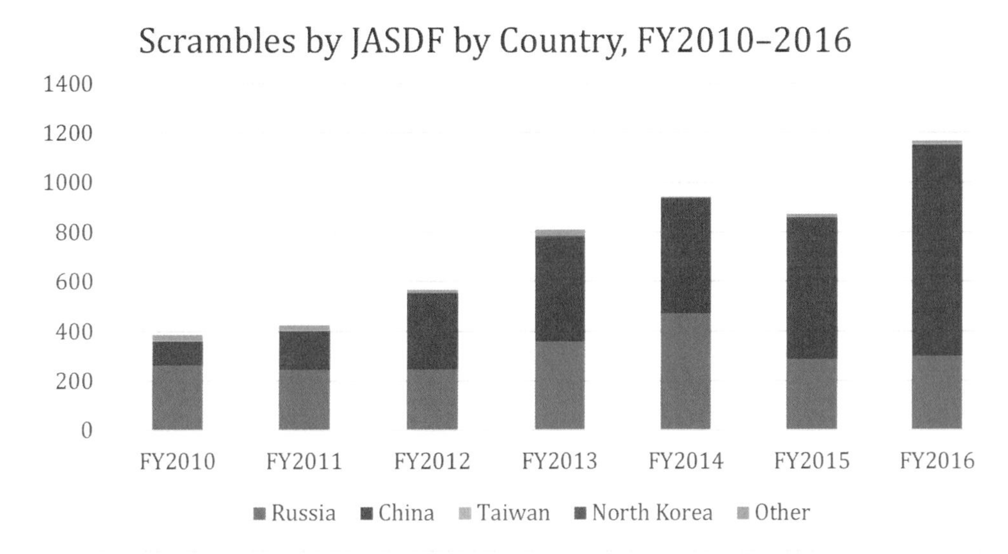

SOURCE: Ministry of Defense of Japan website.

Figure 4.6. Total Scrambles by the Japanese Air Self-Defense Force Through September 30, 2017

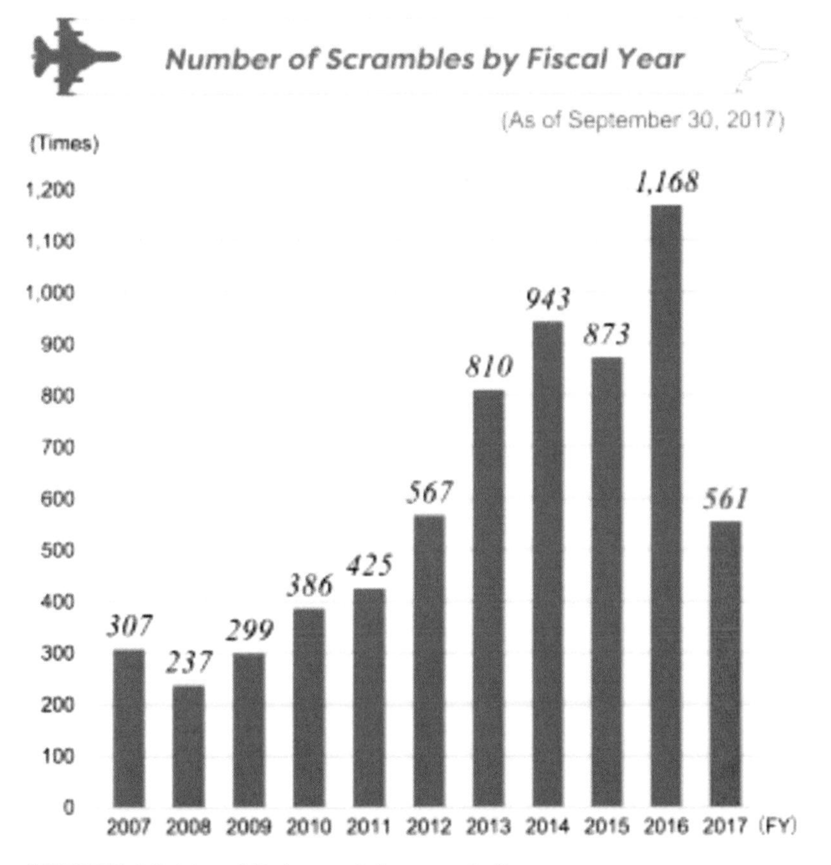

SOURCE: Ministry of Defense of Japan website.

Related Measures Taken by Japan That Support the Aerial Domain

Japan has also expanded its capabilities to monitor air activity and defend against air threats near the Senkaku Islands. In March 2016, the GSDF began operations of a coastal observation unit and logistics facility on Yonaguni to provide constant monitoring of activities in the East China Sea. Manned with about 160 personnel and located 150 km from the Senkaku Islands, the coastal observation unit is a permanent intelligence-gathering facility that provides constant monitoring of activities in the East China Sea. Before its establishment, the closest radar site to the Senkaku Islands was based 200 km away on Miyako. According to Japanese media, the facility on Miyako struggled to detect aircraft flying at low altitudes of several hundred meters due to the curvature of the earth.[34]

Issues and Concerns

While such actions as those described in this chapter are necessary in the face of China's growing air presence, they have the cumulative effect of adding additional pressure to an already overstretched JASDF. On the basis of the total number of scrambles conducted in FY 2016, the JASDF conducted an average of roughly three scrambles per day. The resulting rise in operational tempo and maintenance requirements for aircraft places immense stress on the JASDF's capacity to respond.

First, the increased operational tempo exacerbates maintenance issues. As is the case with other air forces, the aircraft used by the JASDF require routine maintenance. While what is meant by *routine* will differ for part and platform, periodic checks are necessary to ensure continued functioning of these assets. Full inspections and maintenance are performed on the main islands of Japan and are largely conducted at factories. The increased demand on the JASDF means that the frequency with which these aircraft require such inspections and maintenance is increasing. In addition to increased resource requirements for the maintenance, Japan also faces limitations in the factory capability available to handle the work. Not wanting to delay maintenance, but unable to meet the needs, the JASDF instead has relied on rotating aircraft out of the Southwestern Defense Area to one of the other three areas that experience less activity. This allows the factories performing the maintenance to adhere to a predictable maintenance schedule.

Likewise, the increased incursions into Japanese airspace are negatively impacting JASDF pilot training. JASDF personnel are required to log a set number of hours to obtain qualifications for different skills. Because of the finite pilot training time being devoted to relatively mundane interception responses, there is an impact on training for other missions. This includes air-to-air combat, aerial refueling, night missions, and air-to-ground combat. Likewise, pilots are unable

---

[34] Tsuyoshi Takasawa, "ASDF Monitoring Senkakus 24 Hrs a Day; Radar-Equipped Aircraft Deployed in Bid to Prevent Intrusions by Chinese Planes," *Daily Yomiuri*, January 28, 2013.

to improve their skills in flight leader training and mass leader training. Instead, JASDF pilots stationed at Naha are proficient in scramble missions and flying with another jet, but their ability to train beyond these activities is limited. While the real-world experience the pilots are gaining is useful, they are unable to devote this time to the study of other missions, resulting in JASDF pilots being unable to improve their qualifications and skills to become better overall pilots. This carries negative implications for the United States as well, as interoperability with the U.S. Air Force (USAF) could become more difficult for the JASDF as eroding resources provide fewer training assets and opportunities to work together.

As the number of Chinese flights and corresponding Japanese scrambles rises, the risk of an incident such as an accidental collision between Japanese and Chinese aircraft has also increased substantially. In one close encounter in May 2014, Chinese Sukhoi Su-27 fighters came as close as 30 m to a Japanese surveillance aircraft flying near the Senkaku Islands.[35]

In another incident, China's Ministry of National Defense accused Japanese fighters of approaching two Sukhoi Su-30s at high speed and locking on to the jets, an allegation denied by Japan's Ministry of Defense.[36] Given the state of tensions in the East China Sea, an accidental collision could quickly spiral into an uncontrollable crisis between the two countries.

This air activity is part of a larger pattern of Chinese military, maritime law enforcement, maritime militia, and fishing activity that is of growing concern to Tokyo. Indeed, Japan is also alarmed by China's intensification of maritime activities surrounding the Senkaku Islands. In particular, Tokyo fears that its coast guard forces could be stretched thin or, worse, overwhelmed by Chinese assets should China swarm the area using tactics similar to those it employed when it surged fishing vessels in August 2016. In the long term, the growing Chinese presence threatens to establish a new normal that is meant to erode Japan's effective administrative control of the Senkaku Islands.

In the long term, Japan's approach to responding to the higher level of Chinese military air activity is not sustainable. Japanese resources are already stressed. As noted earlier, China's air activities already have spurred Japan to extraordinary measures, such as reorganizing its air defense structures, including establishing new units in all domains; doubling the number of fighters in the sector to respond to Chinese aviators; and increasing its defense spending. However, China's quantitative superiority in terms of available fighter aircraft provides options that could add even more pressure to the JASDF. Nationwide, China has 1,700 fighter aircraft, while Japan has about 288.[37] China can decide at will to escalate its operations near the Senkakus

---

[35] Nobuhiro Kubo, "Japan Protests China Fighter Jet's Close Brush over East China Sea," Reuters, June 11, 2014.

[36] Jesse Johnson and Reiji Yoshida, "Tokyo Denies Beijing's Claim That Japanese Jets Locked Targeting Radars on Fighters over East China Sea," *Japan Times*, July 5, 2016.

[37] Department of Defense of the United States, *Annual Report to Congress: Military and Security Developments Involving the People's Republic of China 2017*, Washington, D.C.: Office of the Secretary of Defense, May 15, 2017, p. 95.

dramatically. Even a minor ramp up in flights per day could strain Japan's ability to respond to the breaking point.

China's approach to Taiwan provides some insight into the scale of the response that Beijing could bring to bear in efforts to stress the air forces of its neighbors, such as Japan. In the case of Taiwan, China increased its fighter presence along the Taiwan Strait center line following then-Taiwanese president Li Teng-hui's July 1999 comments about how Taiwan and China's relations should be on a state-to-state basis.[38] China reportedly patrolled the center line 1,100 times in 1999, 1,500 times in 2001, and 1,700 times in 2005. Taiwan authorities state that, since then, Chinese fighters have flown along the center line an average of 6–12 times per day, and as many as 18–24 times a day.[39] If China chose to escalate the rate of sorties near Japan to anything close to even the 2001 or 2005 levels just mentioned, Japan's relative handful of aviation forces would likely struggle to mount enough aircraft sorties to meet every incursion over a sustained period. Moreover, in such a scenario of extremely high operational tempo, the time available to JASDF pilots to devote to important types of training would erode further. If China were to escalate its operations even more dramatically, it would likely aim to force Japan to ever-costlier operational responses, which China has the mass to sustain almost indefinitely, while Japan clearly does not. The goal, then, would be to exert sustained military pressure through "routine" operations that exact a political toll on the Abe government in the near term while also asserting China's commitment to a sustained military presence over the long term.

---

[38] Taipei Central News Agency, "China's M503 Route Allows Taiwan Leeway for Defense: NSB Chief," March 26, 2015.

[39] Taipei Central News Agency, 2015.

5. Conclusion

China seeks to overtake Japan as the dominant power in the region. As part of that effort, China is intent on challenging Japan's administrative control over the Senkaku Islands and on demonstrating that it can exercise control in the area while avoiding escalation to a military conflict with Japan. At the same time, China's East China Sea air activities also provide its aviators with invaluable experience that serves to expose Chinese pilots to new operational environments and train them to be able to execute offensive missions. Additionally, these activities may deter Japan and could elicit responses that permit China to depict Japan as the aggressor.

As the confrontation over the Senkakus in particular gains political and strategic significance, both countries face a strong incentive to maintain robust operational activity and to avoid compromises that could signal a weakening of resolve. While an unexpected crisis escalating to war is not in the interest of either country, the concentrated presence of military and paramilitary air and maritime platforms increases the risk of a miscalculation occurring. Careful management of this potential flashpoint will test the leadership and skills of Japan, the United States, and China in the coming years.

To counter China, Japan has largely prioritized posture changes and increased the quantity of aircraft, radar, and other key assets in the region. While significant, it is unlikely that this response is sustainable over the long term, given Japan's resource constraints and China's growing numerical superiority in aircraft and naval platforms. Although SDF and Japan Coast Guard budgets have increased, they have only grown some 5 percent in absolute terms since 2012, and just under half of the SDF budget continues to be devoted to personnel expenditures, so the air fleet is not expected to increase significantly. Indeed, given anticipated retirements of the F-2 and F-15 airframes, Japan could face a reduced inventory of fighter aircraft, even as it awaits delivery of a limited number of F-35s.[1] Therefore, in-domain responses that do not further tax pilots or fighter airframes might be useful.

More broadly, a major increase in defense spending, and thus a larger force structure, would help address the imbalance of Chinese-Japanese forces, but Japan is restrained by normative domestic restraints. Although not legally binding or documented as policy, with the exception of symbolic breaks in the 1980s, Japanese governments have constrained their defense spending to no more than 1 percent of GDP. This self-imposed restraint was instituted in 1976 in response to a mix of continuing antiwar sentiment among the public, economic recovery needs, and considerations of appropriate peacetime defense spending. Over time, it became a barometer

[1] Paul Kallender-Umezu, "Japan's Fighter Procurement Crunch," *Defense News*, June 6, 2015.

of Japanese "normalization," and is thus politically difficult to break. Yet, despite Prime Minister Shinzo Abe saying last year that he would break the barrier, the FY 2018 defense budget still hovers at 1 percent of GDP. Even if Tokyo could dramatically increase spending, with its defense budget remaining roughly one-third of China's, there is little possibility that Japan can address the imbalance of forces through increased spending without causing significant political upheaval at home.

Because the situation around the Senkaku Islands is an issue of Japanese domestic air sovereignty, the United States has little room to assist Japan directly in countering Chinese air activity in the East China Sea. If, however, a situation escalates into a conflict that involves military force, then there is greater latitude for the United States to provide assistance. Short of a more forward-leaning approach that escalates horizontally into other domains where the alliance has advantages, the USAF is limited in what it can do directly to assist Japan in countering Chinese activity. One possible indirect way to achieve greater USAF involvement would be operational actions that force China to divert air assets away from Japan and the East China Sea. For example, if the USAF significantly increased its reconnaissance flights throughout the East China Sea area, Chinese responses could draw fighter aircraft away temporarily and provide Japan some breathing space. Operations in China's Eastern Theater would have the best chance of straining the PLA, so these orbits would have to be in that area as opposed to in the South China Sea.

Given limited U.S. options to assist Japan directly in responding to Chinese military air activities, the following recommendations involve possible discussions between the USAF and Department of Defense with their Japanese counterparts to enhance Japan's management of the issue. While Japan's air defense is solely the responsibility of the JASDF, the 2015 Guidelines for U.S.-Japan Defense Cooperation explicitly emphasize that the aim for the alliance is, among other things, "seamless, robust, flexible, and effective bilateral responses" and "synergy across the two governments' national security policies."[2] To achieve this, the allies should be in sync about the challenges posed by China and exchange best practices on how to prepare and respond to them. The following recommendations are a few suggestions for achieving this.

First, U.S. and Japanese officials should exchange views on ways that Japan could respond quickly and effectively to any surge scenarios involving sudden, large numbers of Chinese military aircraft flight operations near Japan. China has already demonstrated its preference for tactics that overwhelm rival militaries with swarms of vessels in the water, as seen in its activities in the Senkakus, or sorties of fighter aircraft, as demonstrated in its actions against Taiwan. Deliberate planning that might require adjustments in sortie protocols or rapid redeployment of fighter aircraft from other parts of the country is solely the responsibility of

[2] Department of Defense of the United States, *The Guidelines for U.S.-Japan Defense Cooperation*, April 27, 2015, p. 1.

the JASDF. But thinking through this issue as an alliance would enable U.S. officials to help frame an effective response that might also deter the PLA from more adventurous and aggressive deployments in the future.

Second, the allies should include the issue of Japanese reprioritization of assets to the southwestern region in their discussions of U.S. force realignment, particularly given the planned move of U.S. Marines from Okinawa to Guam over the next decade.[3] To meet the rapid uptick in Chinese activity, Japan has not only boosted various platforms specializing in island defense but also focused on bolstering the GSDF presence on the islands between Okinawa and Yonaguni. With the exception of one JASDF warning radar unit on Miyako, the JASDF and MSDF have no presence throughout the southwestern region despite being tasked with its defense. Given the expected U.S. force reductions in the coming decade, the allies should discuss the possible deployment of air and naval units and forces to the southwestern islands, assuming there is sufficient physical space to build them. Because the JASDF will procure the Global Hawk from the United States, one possible step could be a permanent operating base once the aircraft enters into service. Another possible tactic could be to relocate some U.S. personnel from other parts of Japan to share joint bases with the SDF on these islands, something the USAF and JASDF do at Misawa Air Base in northern Japan.

Understanding that the establishment of the 9th Air Wing was a historic change for the JASDF, and directly connected to increasing Chinese activity, U.S. officials could share relevant Cold War experiences with analogues of how changing operational situations necessitated the moving of assets to counter Soviet incursions. This could serve as background for an assessment of the feasibility and effectiveness of adding a new air wing in Okinawa or further reliance on assets from other parts of Japan, when necessary. For example, given that the airspace patrolled by the Southwestern Air Defense Force now accounts for the majority of the JASDF's scrambles each year, a second air wing in the southwest of the country would help relieve some pressure on the 9th Air Wing but would be expected to incur costs in resources, space, and personnel.

Third, U.S. officials can share experiences of how scrambling protocols evolved during the Cold War to meet the changing situation. It is already known that the volume of Chinese air activity forced the JASDF to depart from its long-standing operational protocol of scrambling two jets for each potential airspace violation and increase it to four aircraft.[4] Given the strain that increased Chinese activity has already imposed on aircraft maintenance schedules and pilot training, it is questionable whether the increase to four jets is sustainable without a significant increase in resources. If it is not sustainable, an increase in Chinese air presence will pose a significant challenge to Japanese control of its airspace and support China's narrative that it can

---

[3] Jeffrey Hornung, *The U.S. Military Laydown on Guam: Progress amid Challenges*, Washington, D.C.: Sasakawa USA, April 2017.

[4] "Japan Doubles Fighter Jets Deployed for Scrambles Against China," *Japan Times*, February 26, 2017.

and does administer the region, thereby possibly becoming an alliance issue. If more resources are required, this too will impact the alliance, given that resources devoted to scrambles may result in fewer combined training opportunities. It is the JASDF's responsibility to decide whether to keep its current scramble protocol or return to a two-jet response protocol, but given the decades-long experience of the USAF vis-à-vis Soviet fighters, it would be helpful to discuss the risks that both protocols could present and any possible means by which to conserve scarce resources without weakening Japan's airspace control or the operational preparedness of the alliance. Another alternative would be for the JASDF to acquire and field unmanned aerial vehicles (UAVs), either armed or unarmed, to respond to Chinese incursions. While such aircraft would not be able to match the speed of fighters in a response, UAVs could become a part of a response package—especially to slower flying transports or reconnaissance missions—that at least gives Chinese aviators pause.

Based on its own operational experiences, the United States should work with Japan to train in how to rely on existing and planned ground-based air defenses as a suitable and appropriate counter to some Chinese air incursions. Not only would this curtail JASDF responses to a narrower set of flight profiles, thereby economizing and reducing strain on its fighter aircraft, but learning from U.S. best practices would also enable Japan to train in this area quickly and more efficiently. This could help Japan determine where to deploy ground-based air defense systems capable of tracking and monitoring Chinese military air incursions, thereby ensuring more efficient reallocation of responsibilities to meet Chinese military aircraft between ground-based air defense and aviation units. As this is solely a sovereign defense issue, Japan would be responsible for developing its own rules of engagement and plans for seams in coverage.

Japan may also want to consider cross-domain and bilateral responses with other nations in its efforts to counter Chinese intransigence. While outside the immediate area of concern and potentially politically difficult, Japan could seek to expand its bilateral security relationships with Southeast Asian nations to include more of its JASDF forces operating in conjunction with partner air and naval forces. Japan's efforts to extend its maritime security relationships with countries such as Indonesia and Vietnam[5] could be complemented with joint air patrols in the South China Sea as part of exercises, for instance. Complementing these maneuvers with an information campaign designed to remind China of its own vulnerabilities in its maritime periphery might warn China that in-domain escalation is an option not only for Beijing but also for the JASDF. In addition, beyond the maritime realm, Japan could consider a training activity or exercise with Indian air forces assigned responsibility for disputed areas of the China-India border. While no doubt provocative, this type of response may be what is required to convey to Beijing the seriousness of its own provocations in the East China Sea. Further cross-domain responses could incorporate Japan's superior naval assets by demonstrating either the intent or

---

[5] Michael Hart, "Japan's Maritime Diplomacy Mission in Southeast Asia," *Diplomat*, August 28, 2017.

ability to close the Miyako Strait to the Chinese navy by training on minelaying in contiguous waters or by basing minesweeping vessels and minelaying aircraft nearby.

Finally, if indeed the Chinese leadership directs the PLA to minimize its patrols and engage in potentially escalatory behavior during critical political events like its party congress, then Japan and the United States might consider planning joint military maneuvers and exercises in the East China Sea timed to the Chinese political calendar to cause China to lose face or at least deal with unfavorable international press coverage that could make it appear ineffectual in comparison during such a stand-down. Given China's pattern of behavior on its periphery over the past five years, in particular, clear signals like these of Japan's resolve, and of the U.S. commitment, are necessary if Japan hopes to check Chinese behavior.

References

Anonymous retired Ground Self-Defense Force officer, correspondence with Jeffrey Hornung, September 27, 2017.

Asia Maritime Transparency Initiative, "Smooth Sailing for East China Sea Fishing," November 30, 2017. As of June 25, 2018: https://amti.csis.org/smooth-sailing-east-china-sea/

Beckley, Michael, "The Emerging Military Balance in East Asia: How China's Neighbors Can Check Chinese Naval Expansion," *International Security*, Vol. 42, No. 2, Fall 2017, pp. 78–119.

Burke, Edmund J., and Astrid Stuth Cevallos, *In Line or Out of Order? China's Approach to ADIZ in Theory and Practice*, Santa Monica, Calif.: RAND Corporation, RR-2055-AF, 2017. As of June 25, 2018: https://www.rand.org/pubs/research_reports/RR2055.html

Burke, Matthew, and Chiyomi Sumida, "Japan Bolstering Southwestern Defenses," *Stars and Stripes*, January 29, 2016. As of October 15, 2017: https://www.stripes.com/news/japan-bolstering-southwest-defenses-1.391166

Chase, Michael S., and Jeffrey Engstrom, "China's Military Reforms: An Optimistic Take," *Joint Forces Quarterly*, No. 83, 4th Quarter 2016.

Chase, Michael S., and Cristina L. Garafola, "China's Search for a 'Strategic Air Force,'" *Journal of Strategic Studies*, Vol. 39, No. 1, 2016, pp. 4–28.

"China Passes Japan as Second Largest Economy," *New York Times*, August 15, 2010.

Connable, Ben, Jason H. Campbell, and Dan Madden, *Stretching and Exploiting Thresholds for High-Order War: How Russia, China, and Iran Are Eroding American Influence Using Time-Tested Measures Short of War*, Santa Monica, Calif.: RAND Corporation, RR-1003-A, 2016. As of April 26, 2018: https://www.rand.org/pubs/research_reports/RR1003.html

Crowley, Philip, "Remarks to the Press," U.S. State Department, September 23, 2010.

"Defense White Paper Highlights Threat Posed by China," *Japan Times*, July 21, 2015.

Denyer, Simon, "Tensions Rise Between Washington and Beijing over Man-Made Islands," *Washington Post*, May 13, 2015. As of November 13, 2017: https://www.washingtonpost.com/world/asia_pacific/tensions-rise-between-washington-and-beijing-over-man-made-islands/2015/05/13/e88b5de6-f8bd-11e4-a47c-e56f4db884ed_story.html

Department of Defense of the United States, *The Guidelines for U.S.-Japan Defense Cooperation*, April 27, 2015. As of June 25, 2018: http://archive.defense.gov/pubs/20150427_--_guidelines_for_us-japan_defense_cooperation_final&clean.pdf

———, *Annual Report to Congress: Military and Security Developments Involving the People's Republic of China 2017*, Washington, D.C.: Office of the Secretary of Defense, May 15, 2017. As of June 25, 2018: https://www.defense.gov/Portals/1/Documents/pubs/2017_China_Military_Power_Report.PDF

Donald, David, "Japan Bolsters Southern Air Defenses," *Aviation International News*, April 22, 2014. As of October 28, 2017: https://www.ainonline.com/aviation-news/defense/2014-04-22/japan-bolsters-southern-air-defenses

"Establishment of the New 9th Air Wing," *Japan Defense Focus*, March 2016.

"Establishment of the Southwestern Defense Force," *Japan Defense Focus*, August 2017. As of October 15, 2017: http://www.mod.go.jp/e/jdf/no91/index.html

Gady, Franz-Stefan, "Japan's Air Force to Receive 100 New Stealth Fighter Jets," *Diplomat*, July 7, 2016: http://thediplomat.com/2016/07/japans-air-force-to-receive-100-new-stealth-fighter-jets/

Hart, Michael, "Japan's Maritime Diplomacy Mission in Southeast Asia," *Diplomat*, August 28, 2017. As of June 25, 2018: https://thediplomat.com/2017/08/japans-maritime-diplomacy-mission-in-southeast-asia/

Holmes, James, and Toshi Yoshihara, "Deterring China in the 'Gray Zone': Lessons of the South China Sea for U.S. Alliances," *Orbis*, Vol. 61, No. 3, Summer 2017, pp. 322–339.

Hornung, Jeffrey, "Japan's Growing Hard Hedge Against China," *Asian Security*, Vol. 10, No. 2, 2014, pp. 97–122.

———, *The U.S. Military Laydown on Guam: Progress amid Challenges*, Washington, D.C.: Sasakawa USA, April 2017. As of June 25, 2018: https://spfusa.org/wp-content/uploads/2017/04/The-U.S.-Military-Laydown-On-Guam.pdf

Isobe, Koichi, "The Amphibious Operations Brigade," *Marine Corps Gazette*, Vol. 101, No. 2, February 2017, pp. 24–29. As of June 25, 2018: https://www.mca-marines.org/gazette/2017/02/amphibious-operations-brigade

"Japan: ASDF, Chinese Fighter Jets Waging Hidden Battles over East China Sea," *Sentaku*, June 1, 2017.

"Japan Coast Guard to Spend 27% of Budget on Boosting Senkaku Surveillance in 2017," *Kyodo News*, December 22, 2016.

"Japan Doubles Fighter Jets Deployed for Scrambles Against China," *Japan Times*, February 26, 2017. As of November 15, 2017: https://www.japantimes.co.jp/news/2017/02/26/national/politics-diplomacy/japan-doubles-fighter-jets-deployed-scrambles-china/#.Wg4E6baZMfg

"Japan Frees Chinese Boat Captain amid Diplomatic Row," *BBC*, September 24, 2010. As of November 13, 2017: http://www.bbc.com/news/world-11403241

Johnson, Jesse, and Reiji Yoshida. "Tokyo Denies Beijing's Claim That Japanese Jets Locked Targeting Radars on Fighters over East China Sea," *Japan Times*, July 5, 2016. As of October 28, 2017: https://www.japantimes.co.jp/news/2016/07/05/national/beijing-accuses-asdf-jets-radar-lock-targeting-fighters-east-china-sea/#.WfkctBOPJTY

Kallender-Umezu, Paul, "Japan's Fighter Procurement Crunch," *Defense News*, June 6, 2015. As of November 12, 2017: https://www.defensenews.com/air/2015/06/06/japan-s-fighter-procurement-crunch/

Kubo, Nobuhiro, "Japan Protests China Fighter Jet's Close Brush over East China Sea," Reuters, June 11, 2014. As of October 28, 2017: http://www.reuters.com/article/us-japan-defense-protest/japan-protests-china-fighter-jets-close-brush-over-east-china-sea-idUSKBN0EM0RJ20140611

Lah, Kyung, "Japan Hopes for U.S. Help in Row with China," *CNN*, November 13, 2010. As of November 13, 2017: http://www.cnn.com/2010/WORLD/asiapcf/11/13/japan.apec.china/index.html

Lim, Louisa, and Frank Langfitt, "China's Assertive Behavior Makes Neighbors Nervous," *All Things Considered*, NPR, November 2, 2012. As of June 25, 2018: http://www.npr.org/2012/11/02/163659224/chinas-assertive-behavior-makes-neighbors-wary

McCurry, Justin, and Tania Branigan, "Obama Says US Will Defend Japan in Island Dispute with China," *Guardian*, April 24, 2014. As of November 13, 2017: https://www.theguardian.com/world/2014/apr/24/obama-in-japan-backs-status-quo-in-island-dispute-with-china

McGregor, Richard, *Asia's Reckoning: China, Japan, and the Fate of U.S. Power in the Pacific Century*, New York: Viking, 2017.

Ministry of Defense of Japan, "Bōuei Daijin Rinji Kisha Kaiken Gaiyō" ["Summary of the Minister of Defense's Special Press Conference"], n.d.a. As of September 7, 2016: http://www.mod.go.jp/j/press/kisha/2016/09/07.html

———, "China's Activities Surrounding Japan's Airspace," n.d.b. As of August 29, 2017: http://www.mod.go.jp/e/d_act/ryouku/

———, "National Defense Program Guidelines for FY2011 and Beyond," December 17, 2010. As of November 13, 2017: http://www.mod.go.jp/e/d_act/d_policy/pdf/guidelinesFY2011.pdf

———, *Medium Term Defense Program (FY2014–FY2018)*, December 17, 2013. As of June 25, 2018: http://www.mod.go.jp/j/approach/agenda/guideline/2014/pdf/Defense_Program.pdf

———, *Defense Programs and Budget of Japan FY2017*, Tokyo: Ministry of Defense, August 2016. As of November 13, 2017: http://www.mod.go.jp/e/d_budget/pdf/290328.pdf

———, "Overview of Situation Surrounding Japan: Statistics on Scrambles Through FY 2016," slide handout, Center for Air Power Strategic Studies, May 8, 2017a.

———, "Reference 10," in *Nihon no Bōei Heisei 29 Nenban* [*Defense of Japan 2017*], Tokyo: Ministry of Defense, 2017b, p. 480.

Ministry of Defense of Japan, Joint Staff, "Statistics on Scrambles Through Fiscal Year 2013," press release, April 23, 2014. As of October 28, 2017: http://www.mod.go.jp/js/Press/press2014/press_pdf/p20140423_02.pdf

———, "On the Flight of Chinese Aircraft in the East China Sea," press release, May 21, 2015a. As of October 28, 2017: http://www.mod.go.jp/js/Press/press2015/press_pdf/p20150521_02.pdf

———, "Statistics on Scrambles Through Fiscal Year 2014," press release, May 22, 2015b. As of October 28, 2017: http://www.mod.go.jp/js/Press/press2015/press_pdf/p20150522_01.pdf

———, Statistics on Scrambles Through Fiscal Year 2015," press release, April 22, 2016. As of October 28, 2017: http://www.mod.go.jp/js/Press/press2016/press_pdf/p20160422_03.pdf

———, "Statistics on Scrambles Through Fiscal Year 2016," press release, April 13, 2017a. As of November 13, 2017: http://www.mod.go.jp/js/Press/press2017/press_pdf/p20170413_02.pdf

———, "Statistics on Scrambles Through First Quarter of Fiscal Year 2017," press release, July 14, 2017b. As of October 28, 2017: http://www.mod.go.jp/js/Press/press2017/press_pdf/p20170714_06.pdf

———, "Statistics on Scrambles During the First Half of FY2017," press release, October 13, 2017c. As of November 26, 2017: http://www.mod.go.jp/js/Press/press2017/press_pdf/p20171027_01.pdf

Ministry of Foreign Affairs of Japan, "Statement by the Minister for Foreign Affairs on the Announcement on the 'East China Sea Air Defense Identification Zone' by the Ministry of National Defense of the People's Republic of China," November 24, 2013. As of November 11, 2017: http://www.mofa.go.jp/press/release/press4e_000098.html

———, "Senkaku Islands," April 15, 2014. As of June 25, 2018: http://www.mofa.go.jp/region/asia-paci/senkaku/index.html

———, "Address by Prime Minister Shinzo Abe at the 'Shared Values and Democracy in Asia' Symposium," January 19, 2016. As of November 13, 2017: http://www.mofa.go.jp/s_sa/sw/page3e_000452.html

———, "Trends in Chinese Government and Other Vessels in the Waters Surrounding the Senkaku Islands, and Japan's Response," June 30, 2017. As of November 13, 2017: http://www.mofa.go.jp/region/page23e_000021.html

Morris, Lyle J., "The New 'Normal' in the East China Sea," *Diplomat*, February 24, 2017a. As of October 28, 2017: https://thediplomat.com/2017/02/the-new-normal-in-the-east-china-sea

———, "Blunt Defenders of Sovereignty: The Rise of Coast Guards in East and Southeast Asia," *Naval War College Review*, Vol. 70, No. 2, Spring 2017b, pp. 75–112.

Panda, Ankit, "East China Sea: Japan Coast Guard Plans Miyako Island Facility Upgrades," *Diplomat*, September 24, 2017. As of June 25, 2018: https://thediplomat.com/2017/09/east-china-sea-japan-coast-guard-plans-miyako-island-facility-upgrades/

Prime Minister of Japan's Office, "National Security Strategy," December 17, 2013. As of November 13, 2017: http://japan.kantei.go.jp/96_abe/documents/2013/__icsFiles/afieldfile/2013/12/17/NSS.pdf

Reeves, Jeffrey, Jeffrey Hornung, and Kerry Lynn Nankivell, eds., *Chinese-Japanese Competition and the East Asian Security Complex: Vying for Influence*, New York: Routledge, 2017.

Reuters, "Military Strength Comparison: China, Japan, and North Korea," March 28, 2013. As of November 13, 2017: https://blogs.thomsonreuters.com/answerson/military-strength-japan-china-north-korea/

Rinehart, Ian E., and Bart Elias, *China's Air Defense Identification Zone (ADIZ)*, Washington, D.C.: Congressional Research Service, R43894, January 30, 2015.

State Council Information Office, People's Republic of China, "The Diversified Employment of China's Armed Forces," April 16, 2013. As of October 24, 2017: http://en.people.cn/90786/8209362.html

———, "Diaoyu Dao, an Inherent Territory of China," August 23, 2014. As of November 13, 2017: http://english.gov.cn/archive/white_paper/2014/08/23/content_281474983043212.htm

———, "China's Military Strategy," May 26, 2015. As of November 13, 2017: http://www.chinadaily.com.cn/china/2015-05/26/content_20820628.htm

———, "China's Policies on Asia Pacific Cooperation," January 11, 2017. As of November 13, 2017: http://english.gov.cn/archive/white_paper/2017/01/11/content_281475539078636.htm

Stokes, Bruce, "Hostile Neighbors: China vs. Japan," *Pew Research Center*, September 13, 2016. As of November 13, 2017: http://www.pewglobal.org/2016/09/13/hostile-neighbors-china-vs-japan/

Strategic Research Group, "A Study on the People Liberation Army's Capabilities and Increasing Jointness in the East China Sea," Air Staff College research memo, *Air Studies*, No. 2, June 17, 2016. As of June 25, 2018: http://www.mod.go.jp/asdf/meguro/center2/AirPower2nd/125memo4.pdf

Sun, Yun, "Rising Sino-Japanese Competition in Africa," *Brookings Institution*, August 31, 2016. As of November 13, 2017: https://www.brookings.edu/blog/africa-in-focus/2016/08/31/rising-sino-japanese-competition-in-africa/

Taipei Central News Agency, "China's M503 Route Allows Taiwan Leeway for Defense: NSB Chief," March 26, 2015.

Takahashi, Kosuke, "JASDF Forms New AEW Squadron in Okinawa," *Jane's Defence Weekly*, April 14, 2014.

Takasawa, Tsuyoshi, "ASDF Monitoring Senkakus 24 Hrs a Day; Radar-Equipped Aircraft Deployed in Bid to Prevent Intrusions by Chinese Planes," *Daily Yomiuri*, January 28, 2013.

"Tokyo Trying to Draw Attention to Mass China Ship Incursions off Senkakus," *Japan Times*, April 17, 2016. As of November 13, 2017: http://www.japantimes.co.jp/news/2016/08/17/national/tokyo-trying-draw-attention-mass-china-ship-incursions-off-senkakus/#.WV4tMhS9bww

United Nations Convention on the Law of the Sea, Part 2, Territorial Sea and Contiguous Zone, Article 19, Meaning of Passage, December 10, 1982. As of June 25, 2018: http://www.un.org/depts/los/convention_agreements/texts/unclos/part2.htm

Wood, Peter, "Snapshot: Eastern Theater Command," *China Brief*, March 14, 2017.

Xinhua, "Xi Jinping Stresses at the Third Collective Study Session of the Political Bureau to Make Overall Planning for Domestic and International Situations," January 29, 2013a.

———, "Xi Jinping Vows Peaceful Development While Not Waiving Legitimate Rights," January 29, 2013b. As of June 25, 2018: http://en.people.cn/90785/8113230.html

———, "China Air Force in West Pacific Drill," May 21, 2015. As of November 11, 2017: http://news.xinhuanet.com/english/2015-05/21/c_134259412.htm

———, "China Air Force Conducts West Pacific Drill, Patrols ADIZ," September 25, 2016. As of October 28, 2017: http://eng.mod.gov.cn/TopNews/2016-09/25/content_4737135.htm

Zheng, Sarah, Liu Zhen, and Kristin Huang, "China 'Halts Road Building' to End Border Stand-Off," *South China Morning Post*, August 29, 2017. As of October 15, 2017: http://www.scmp.com/news/china/diplomacy-defence/article/2108686/china-halted-road-building-end-india-border-standoff